D

# fractal 3D magic

Introduction by CLIFFORD A. PICKOVER,
Author of *The Math Book*

STERLING
New York

STERLING
New York

An Imprint of Sterling Publishing
387 Park Avenue South
New York, NY 10016

ISBN 978-1-4549-1263-7

Distributed in Canada by Sterling Publishing
c/o Canadian Manda Group, 165 Dufferin Street
Toronto, Ontario, Canada M6K 3H6
Distributed in the United Kingdom by GMC Distribution Services
Castle Place, 166 High Street, Lewes, East Sussex, England BN7 1XU
Distributed in Australia by Capricorn Link (Australia) Pty. Ltd.
P.O. Box 704, Windsor, NSW 2756, Australia

For information about custom editions, special sales, and premium and corporate purchases,
please contact Sterling Special Sales at 800-805-5489 or specialsales@sterlingpublishing.com

# contents

# introduction:
## what is a fractal?

today, computer-generated fractal patterns are everywhere. From squiggly designs on computer art posters to illustrations in the most serious of physics journals, interest continues to grow among scientists and, rather surprisingly, artists and designers. The word fractal was coined in 1975 by mathematician Benoît Mandelbrot to describe an intricate-looking set of curves, many of which were never seen before the advent of computers with their ability to quickly perform massive calculations. Fractals often exhibit self-similarity, which suggests that various exact or inexact copies of an object can be found in the original object at smaller size scales. The detail continues for many magnifications—like an endless nesting of Russian dolls within dolls. Some of these shapes exist only in abstract geometric space, but others can be used as models for complex natural objects such as coastlines and blood vessel branching. These dazzling computer-generated images can be intoxicating, motivating students' interest in math more than any other mathematical discovery in the last century.

Physicists are interested in fractals because they can sometimes describe the chaotic behavior of real-world phenomena such as planetary motion, fluid flow, the diffusion of drugs, the behavior of inter-industry relationships, and the vibration of airplane wings. (Chaotic behavior often produces fractal patterns.) Traditionally, when physicists or mathematicians saw complicated results, they looked for complicated causes. In contrast, many fractal shapes reveal the fantastically complicated behavior of the simplest formulas.

Early explorers of fractal objects include Karl Weierstrass, who in 1872 considered functions that were everywhere continuous but nowhere differentiable, and Helge von Koch, who in 1904 discussed geometric shapes such as the Koch Snowflake. In the nineteenth and early twentieth centuries, several mathematicians explored fractals in the complex plane; however, they could not fully appreciate or visualize these objects without the aid of the computer.[1]

Clifford A. Pickover
Yorktown Heights, NY

# types of fractals

## mandelbrot set

david Darling writes that the Mandelbrot set, or M-set for short, is the "best known fractal and one of the most...beautiful mathematical objects known." *The Guinness Book of World Records* called it "the most complicated object in mathematics." Arthur C. Clarke emphasizes the degree to which the computer is useful for gaining insight: "In principle, [the Mandelbrot set] could have been discovered as soon as men learned to count. But even if they never grew tired, and never made a mistake, all the human beings who have ever existed would not have sufficed to do the elementary arithmetic required to produce a Mandelbrot set of quite modest magnification."

The Mandelbrot set is a fractal, an object that continues to exhibit similar structural details no matter how much the edge of the object is magnified. Think of the beautiful M-set images as being produced by mathematical feedback loops. In fact, the set is produced by iteration, or repetition, of the very simple formula $z_{n+1} = z_n^2 + c$, for complex values of $z$ and $c$, and for $z_0 = 0$. The set contains all points for which the formula does not produce values that diverge to infinity. The first crude pictures of the M-set were drawn in 1978 by Robert Brooks and Peter Matelski, followed by the landmark paper by Mandelbrot in 1980 on its fractal aspects and the wealth of geometric and algebraic information it conveys.

The M-set structure contains super-thin spiral and crinkly paths, connecting an infinite number of island shapes. Computer magnifications of the M-set will easily yield pictures never seen before by human eyes. The incredible vastness of the M-set led authors Tim Wegner and Mark Peterson to remark: "You may have heard of a company that for a fee will name a star after you and record it in a book. Maybe the same thing will soon be done with the Mandelbrot set!"[2]

A number of computer programs now allow us to create stunning artwork based on the Mandelbrot set. FractalWorks,[3] for example, begins with the basic Mandelbrot structure and can use either iteration data or distance estimate data to create striking 3D "height map" images. A histogram of iteration counts can be used to assign colors to plots, generating color schemes, which contribute to the creation of vibrant fractal flowers or alien landscapes, as shown on the following pages.

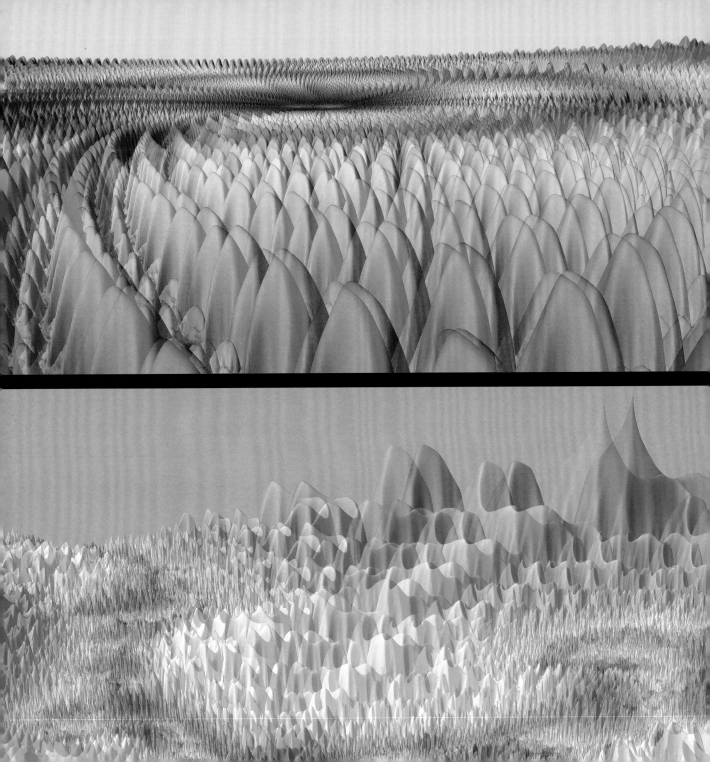

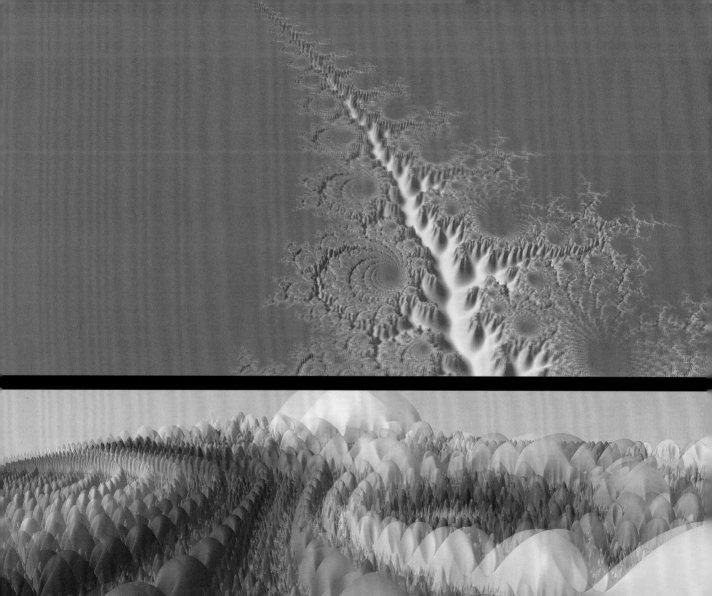

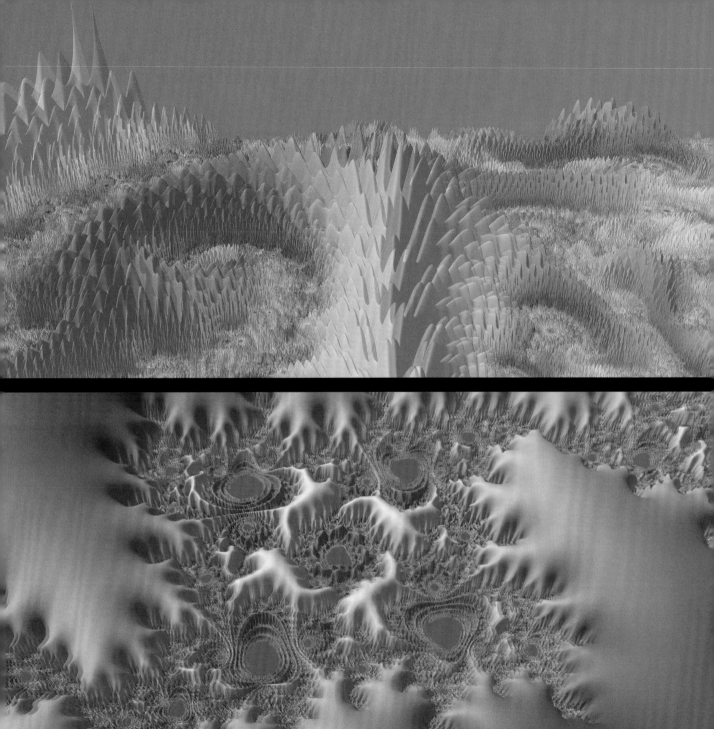

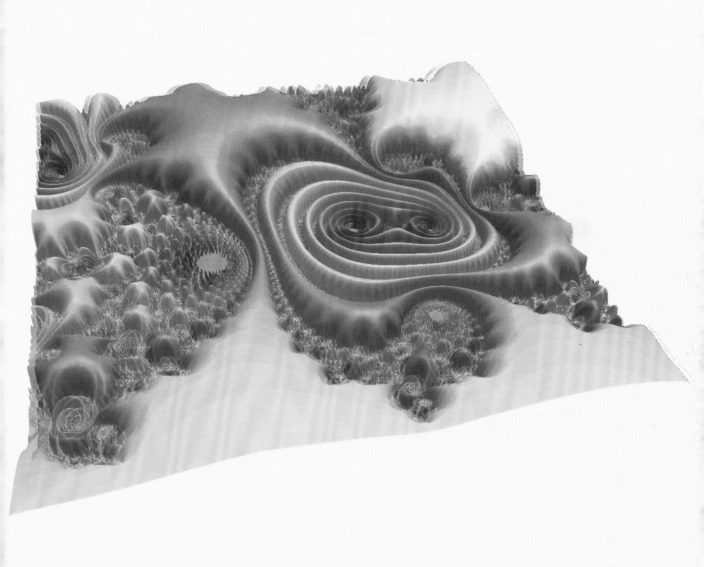

# julia set

The pioneering work of the French mathematician Gaston Julia, published in 1918 when he was twenty-five years old, was essentially forgotten by the mathematical world until the 1980s, when computers enabled the visualization of his mathematics. Julia's idea was to observe the behavior of the orbit of a complex number under iteration of a function $f$. That is, begin with a complex number $z_0$, visualized as a point in the complex plane, and apply $f$ to $z_0$. A complex number $z_0$ is called a prisoner if its orbit under $f$ is bounded, and an escapee if the orbit is unbounded—that is, terms in the orbit move arbitrarily far away from 0. The set of all prisoners for a given function $f$ is called its *prisoner set*, and the set of all escapees is called the *escape set*.[4]

The Julia set can be beautifully visualized as the boundary between the prisoner set and escape set. Like its close relative, the Mandelbrot set, the Julia set is built on the function of the form $f(z) = z_{n+1} = z_n^2 + c$,

where $c$ is some fixed complex number. It has been proven that if the orbit of a starting value $z_0$ ever leaves the circle of radius 2 centered at the origin, then the rest of the terms in the orbit get successively farther from 0, and the orbit is unbounded. Such a starting value $z_0$ would therefore be an element of the escape set for $f$.

Julia sets can be either connected or disconnected (dust-like), depending on the value of the seed point, $c$. If $c$ is taken from a specific spot on the Mandelbrot set, it is considered connected and may appear as a series of closed loops and dendrites. Examples of these include "Douady rabbits," named after French mathematician Adrien Douady, which are shown in the illustrations on pages 26–27. When $c$ lies outside the Mandelbrot set, however, the resulting fractal is disconnected and dust-like, as shown in the illustration on page 28.

As with the Mandelbrot set, Fractal-Works allows us to generate colorful, 3D Julia fractal maps that can resemble everything from flowers to elephants.

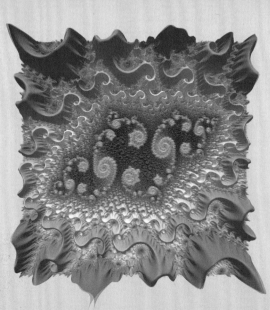

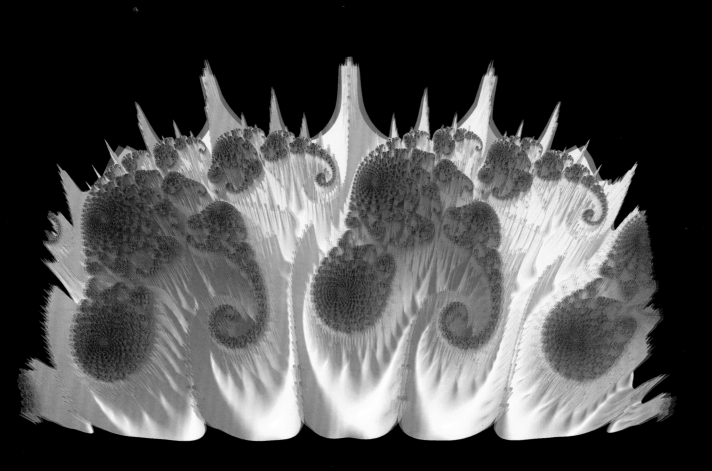

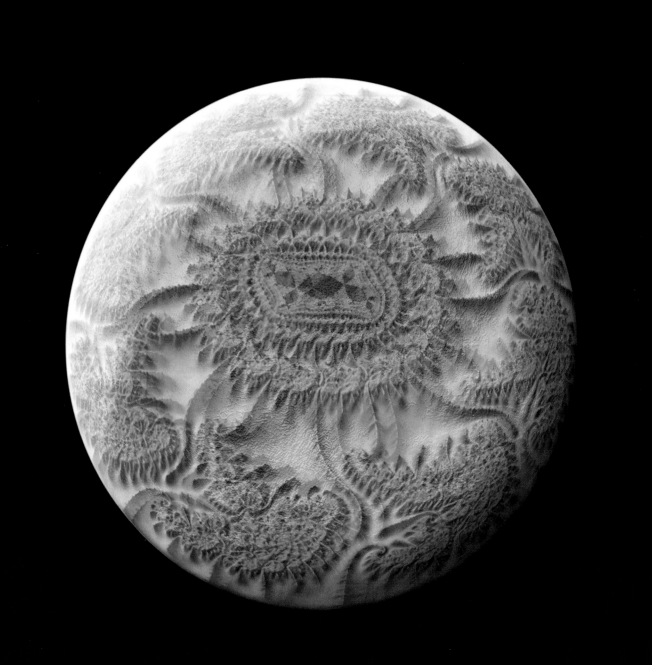

# spirals

Since around 1975, Benoit Mandelbrot and others have offered many descriptions and definitions of fractals. In some interpretations, fractals are thought of as objects that do not lose their detail or their proportions when they are magnified or shrunk, even by many orders of magnitude. This self-similarity is evoked in the endless logarithmic spiral defined by the function $r = ae^{b\Theta}$, where $r$ is the radius, $e$ is the base of natural logarithms, $a$ and $b$ are positive constants, and $\Theta$ is the angle from the $x$-axis. Examples of logarithmic spirals are found in nature, as exemplified in the growth patterns of nautilus shells and Romanesco broccoli, but they also appear as components of Mandelbrot and Julia fractals in many illustrations.

# sponges, carpets, and gaskets

One way to generate fractals is to subdivide a shape into smaller versions of itself and remove one or more copies, continuing recursively. For example, subdividing an equilateral triangle into four equilateral triangles, removing the middle triangle, and recursing generates the Sierpinski triangle—a fractal easily identifiable in 3D "pyramid" form on page 33. In three dimensions, the same process involving cubes produces the Menger sponge. Like the Sierpinski triangle, the Menger sponge is a fractal object with an infinite number of cavities—a nightmarish object for any dentist to contemplate. The object was first described by Austrian mathematician Karl Menger in 1926. To construct the sponge, we begin with a "mother cube" and subdivide it into 27 identical smaller cubes. Next, we remove the cube in the center and the six cubes that share faces with it. This leaves behind 20 cubes. We continue to repeat the process forever. The number of cubes increases by $20^n$, where $n$ is the number of iterations performed on the mother cube. The second iteration gives us 400 cubes, and by the time we get to the sixth iteration, we have 64,000,000 cubes. Illustrations that incorporate such iterations are found on pages 34 and 35.

Each face of the Menger sponge is called a Sierpinski carpet. Fractal antennae based on the Sierpinski carpet are sometimes used as efficient receivers of electromagnetic signals. Both the carpets and the entire cube have fascinating geometrical properties. For example, the sponge has an infinite surface area while enclosing zero volume.

According to the Institute for Figuring, with each iteration, the Sierpinski carpet face "dissolves into a foam whose final structure has no area whatever yet possesses a perimeter that is infinitely long. Like the skeleton of a beast whose flesh has vanished, the concluding form is without substance—it occupies a planar surface, but no longer fills it." This porous remnant

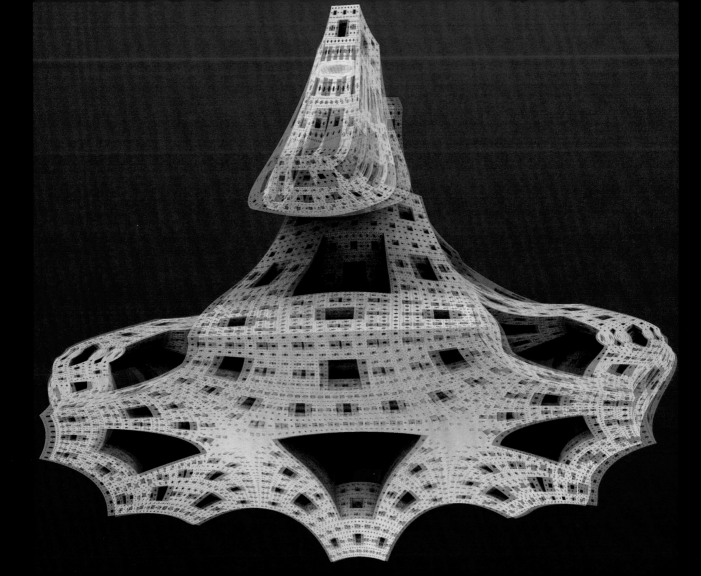

hovers between a line and a plane. Whereas a line is one-dimensional and a plane two-dimensional, the Sierpinski carpet has a "fractional" dimension of 1.89. The Menger sponge has a fractional dimension (technically referred to as the Hausdorff Dimension) between a plane and a solid, approximately 2.73, and it has been used to visualize certain models of a foam-like space-time. Dr. Jeannine Mosely has constructed a Menger sponge model from more than 65,000 business cards that weighs about 150 pounds (70 kilograms).[5]

Another cavity-filled fractal is the Apollonian gasket, named after Greek mathematician Apollonius of Perga. This fractal is generated from triples of circles, where each circle is a tangent to the other two. Patterns that are reminiscent of these gaskets are found in many computer-generated illustrations, such as the futuristic space station–like image on page 36.

# flames

"On the screen they are luminous, twisting, elastic shapes, abstract tangles and loops of glowing filaments," writes Mitchell Whitelaw in *Metacreation: Art and Artificial Life*. These ethereal fractals, which were created by software artist Scott Draves in 1992, can be generated with programs such as the open-source editor and renderer, Apophysis. Like other fractals exhibited in this book, fractal flames are iterated function system (IFS) fractals—that is, they are made up of several possibly overlapping copies (iterations) of themselves, ad infinitum, with each copy being transformed by a function.

Flames differ in a few ways from more familiar IFS fractals, however. For one, instead of iterating linear functions, they iterate nonlinear functions. Secondly, their tone mapping (e.g. contrast and brightness) and coloring are often designed to display as much detail of the fractal as possible, resulting in the "loops of glowing filaments" that reveal the fractal's recursive path.

In addition to inventing the "flame" algorithm, Draves leads the computing project known as Electric Sheep that animates and evolves fractal flames, which are distributed to networked computers and represented as screensavers. The project takes its name from iconic sci-fi author Philip K. Dick's novel *Do Androids Dream of Electric Sheep?*— a fitting metaphor for the process by which computers (androids) begin rendering (dreaming) the fractal movies (sheep). These "sheep" can be generated in a few ways: (1) members of a mailing list can create and submit them; (2) members can download the parameters of existing sheep and tweak them; or (3) a sheep's fractal code can be interpolated or combined ("mated") with that of another sheep automatically by the server or manually by server admins ("shepherds"). Draves and his team of engineers take the metaphor even further: the hundred or so sheep stored on the server at any time are referred to as "the flock," and changes to the code are known as "mutations." In 2011, the project boasted about 500,000 active users.

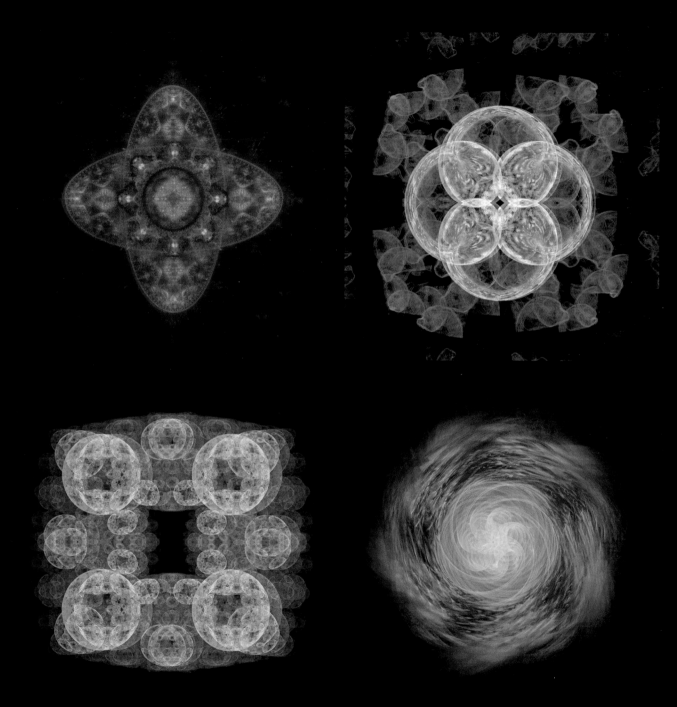

gallery

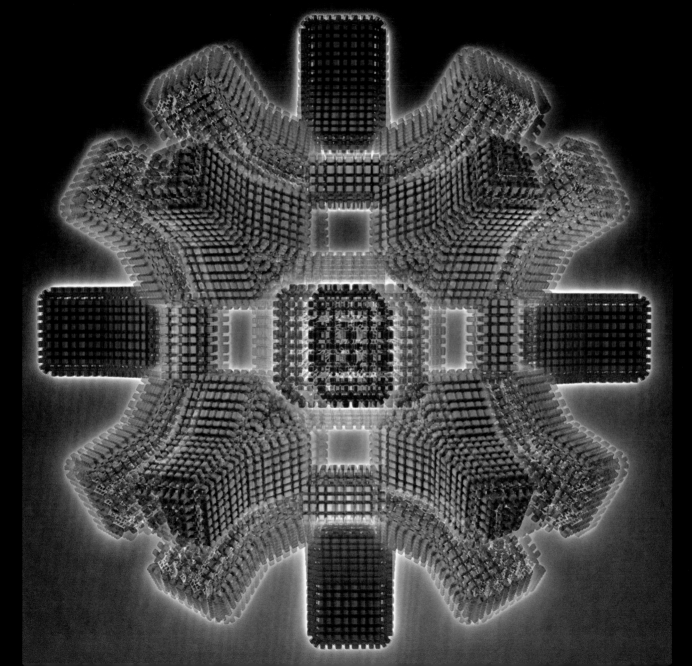

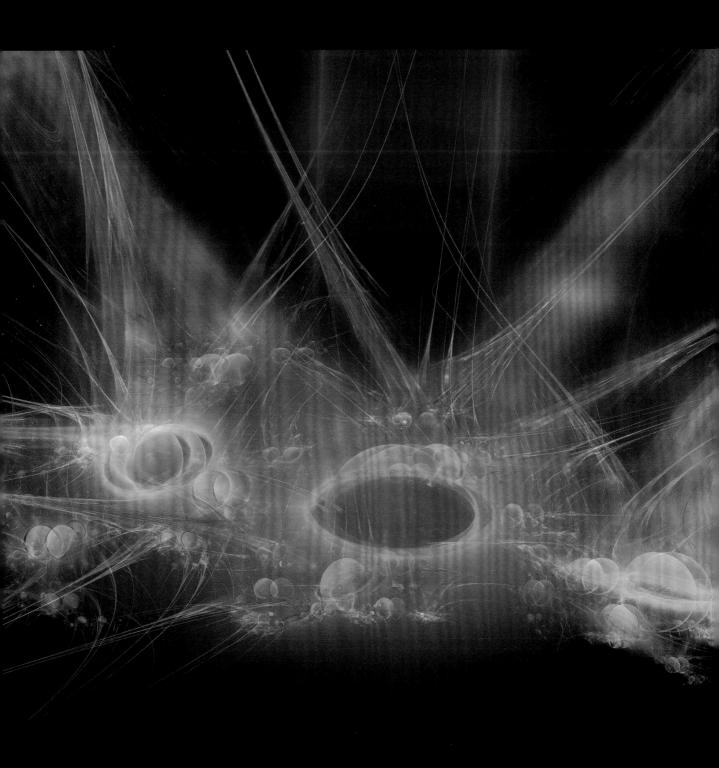

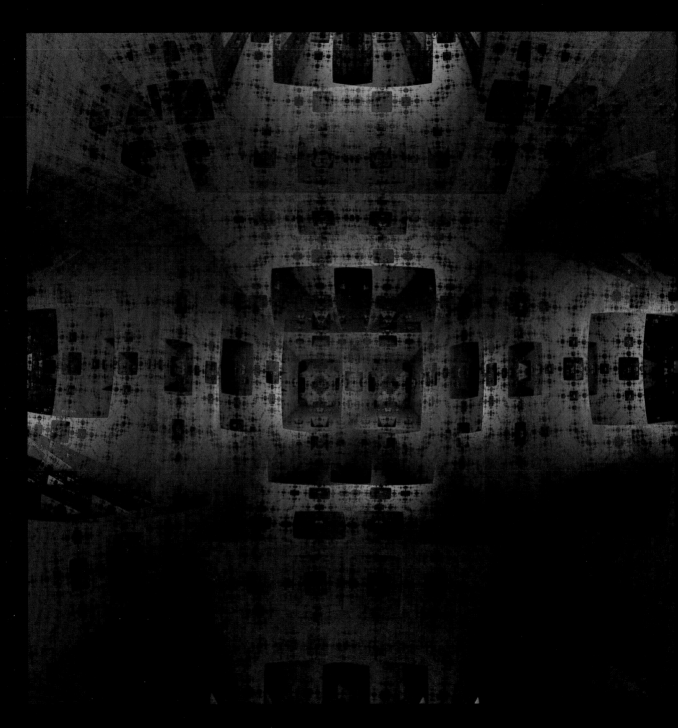

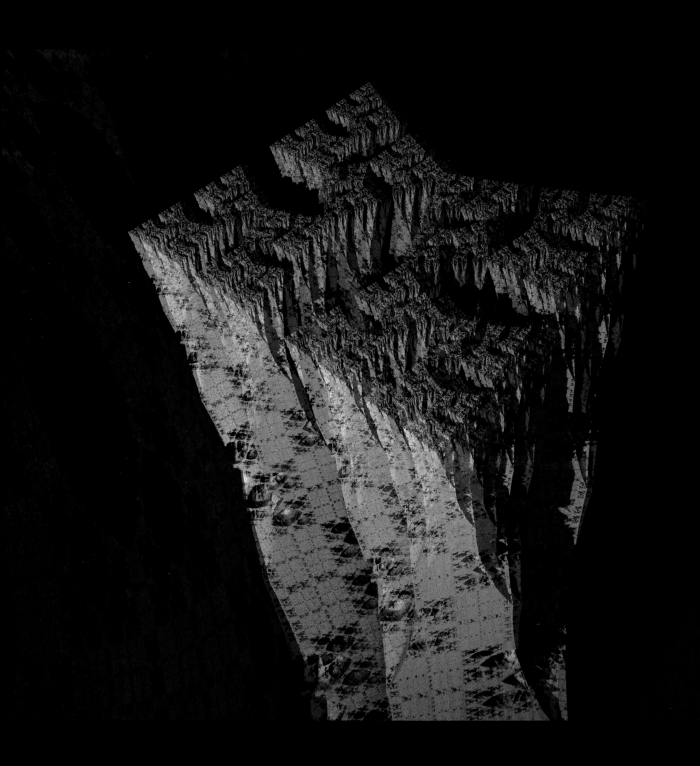

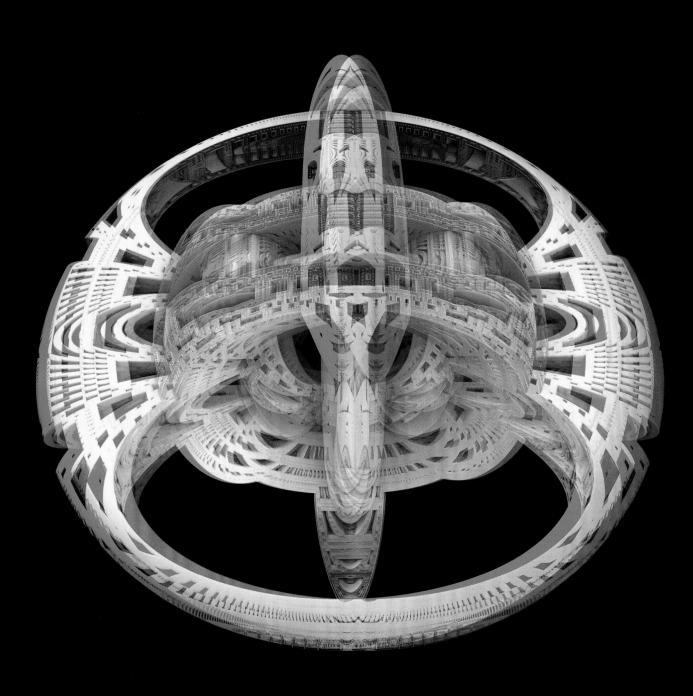

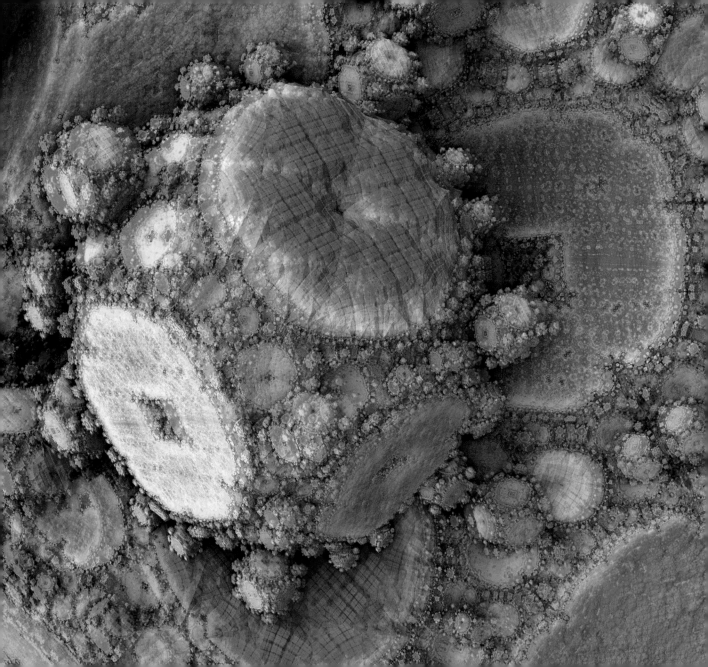

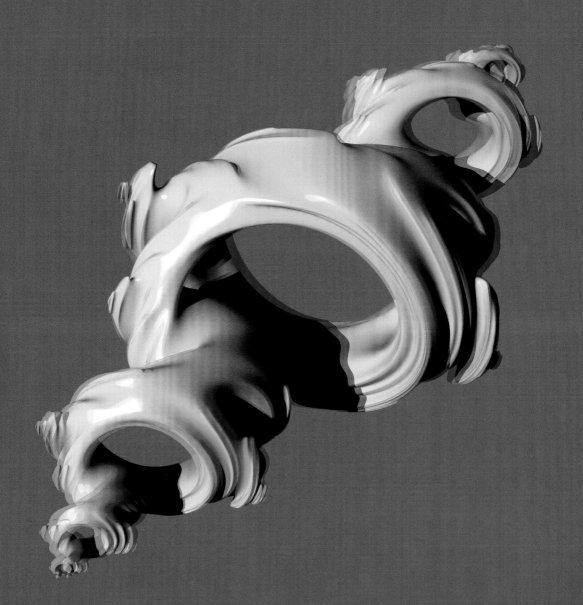

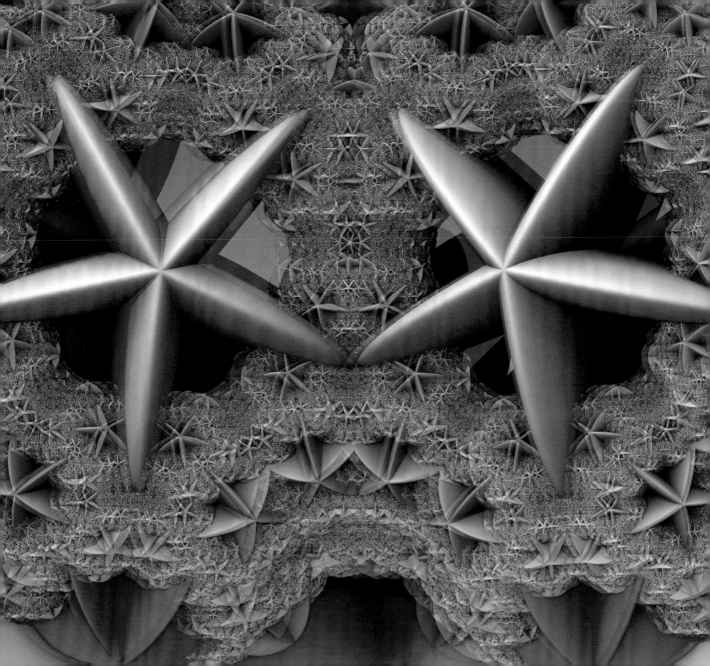

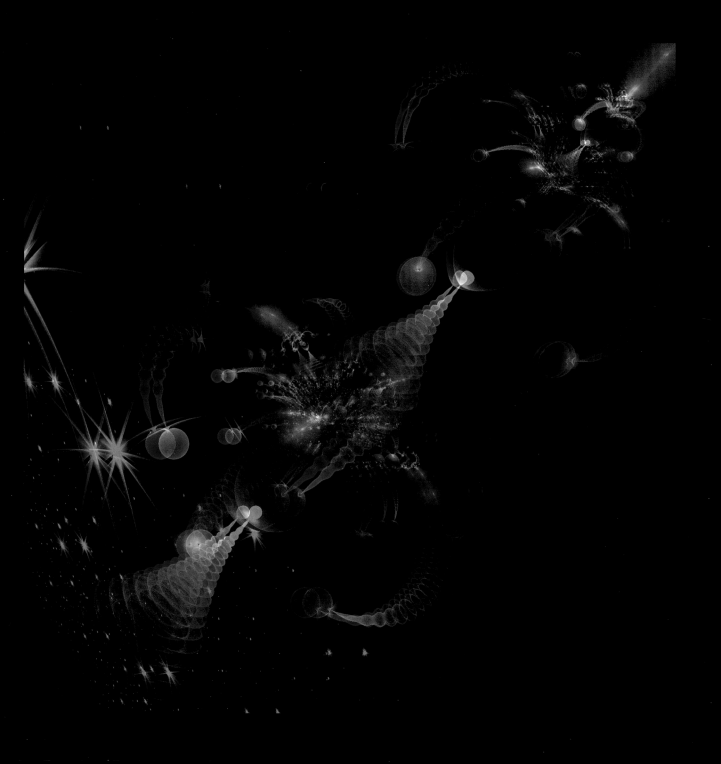

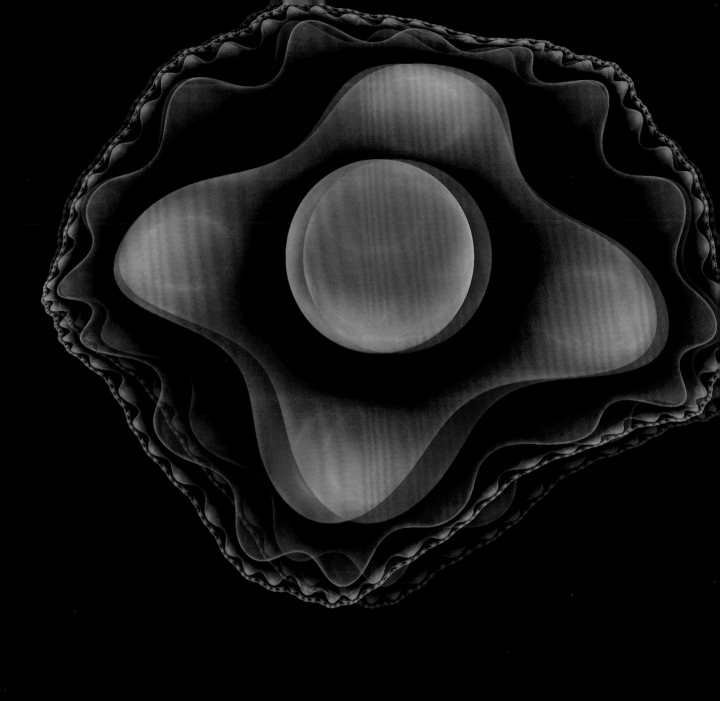

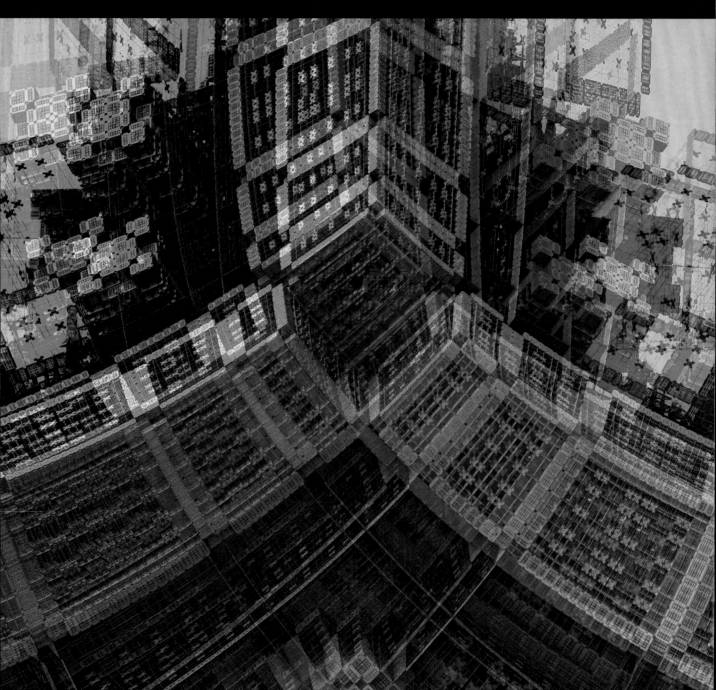

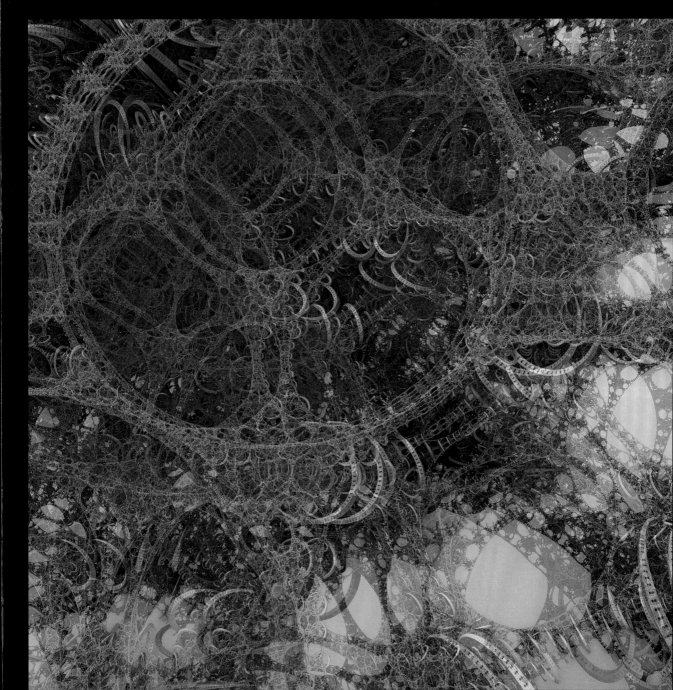

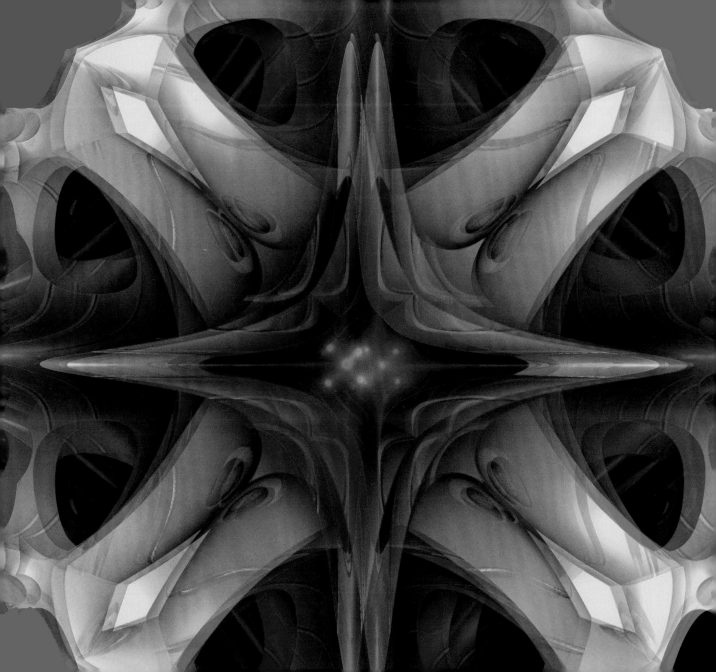

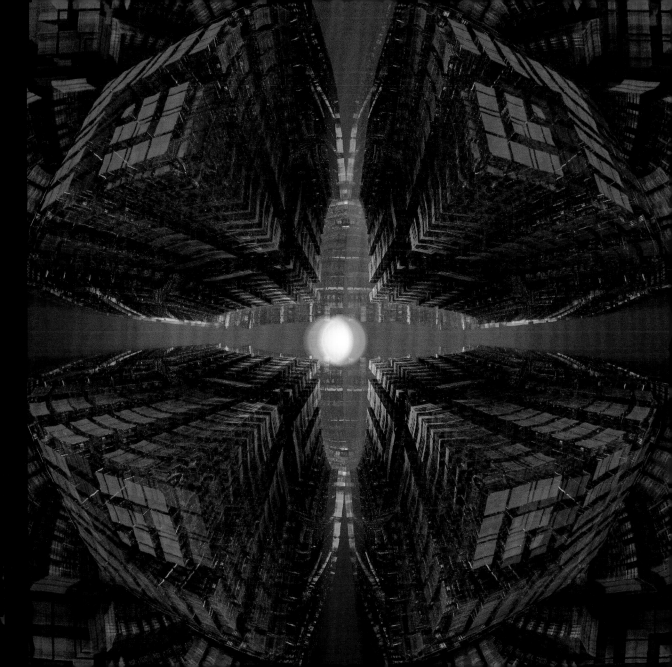

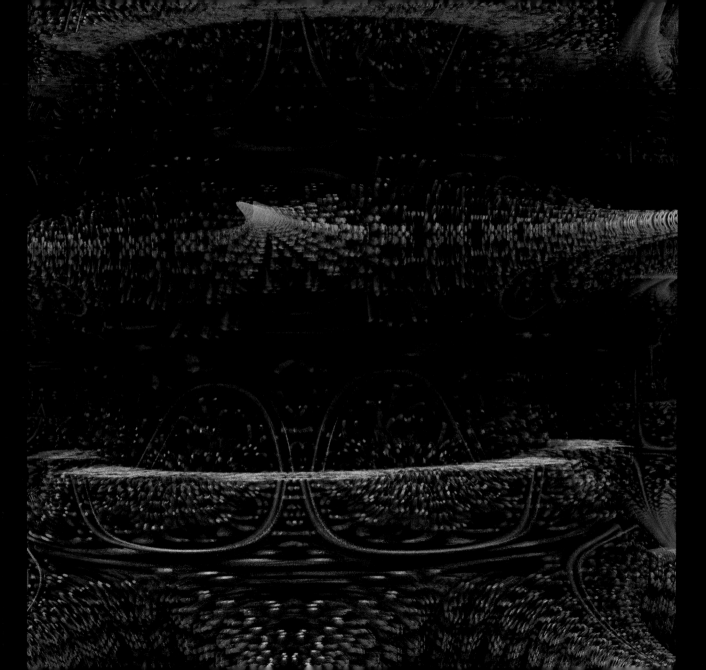

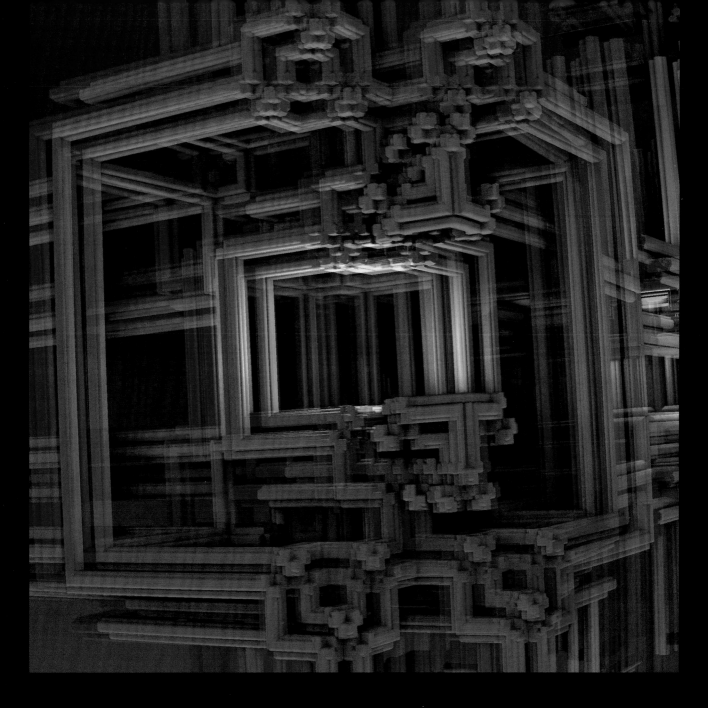

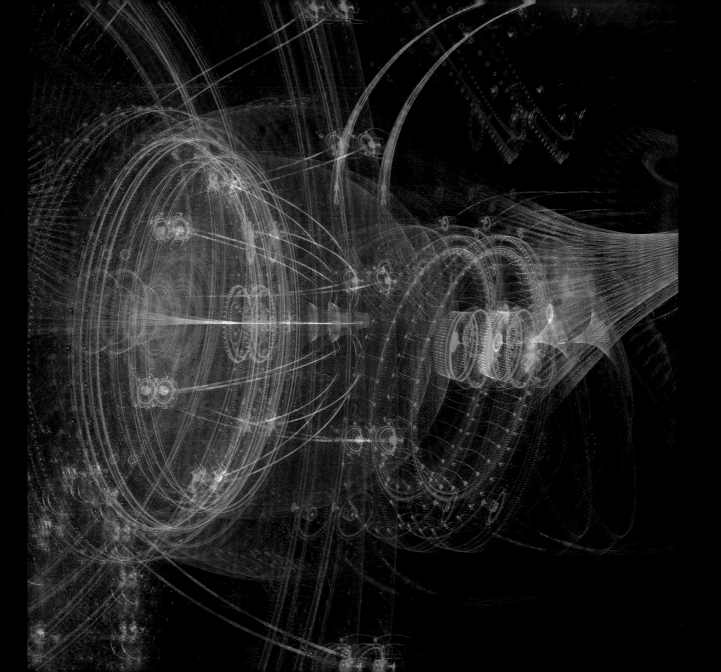

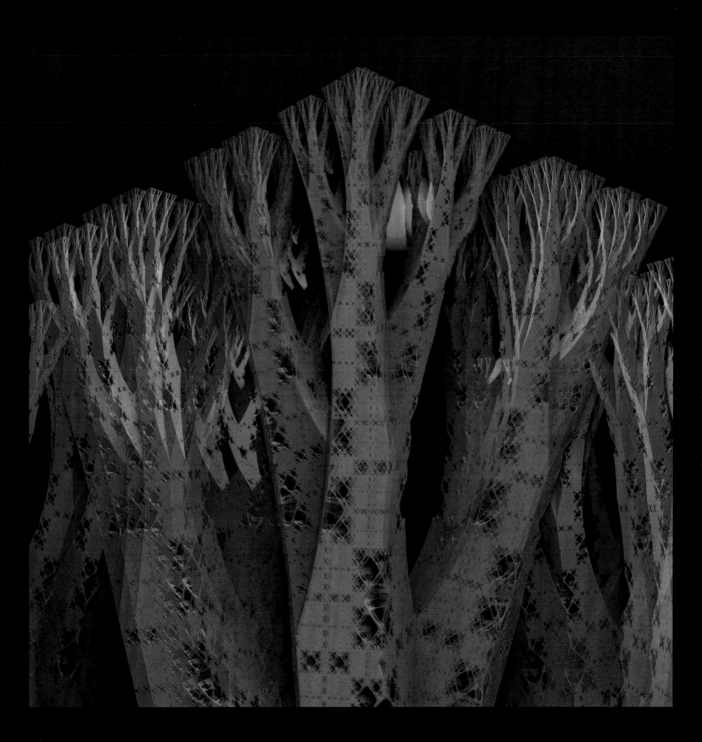

# notes

1. Clifford A. Pickover, *The Math Book* (Sterling, 2009), 460.

2. Ibid., 472.

3. FractalWorks is a high-performance fractal-generating program for the Macintosh platform, created by Duncan Champney. It was first available as shareware in 2007, and has since been released as a commercial product. Inspired by *The Beauty of Fractals* (1986) by Heinz-Otto Peitgen and Peter Richter, FractalWorks focuses on Mandelbrot and Julia sets and offers some unique capabilities: (1) it is multi-threaded, taking advantage of the large number of cores on modern computers; (2) it uses a boundary-following algorithm to greatly speed up the rendering of many plots; (3) it can assign colors to plots using a histogram of iteration counts in an image, which gives good contrast to areas of important detail, and lower contrast to areas that would otherwise display as a riot of colors; (4) it will create 3D "height map" images via either iteration data or distance estimate data, although the latter method generates superior images. Once a height value is assigned to each point, the resulting height map may be rendered as a 3D shape in OpenGL. Colors from the 2D plot are applied to the 3D image, along with realistic directional, global, and specular lighting that the user has full control of. These 3D renderings can be displayed using perspective projection, as stereoscopic images using red/cyan anaglyph images, or as separate left and right eye full-color images. Complete with sample images, including quite a few 3D images that you can use as a starting point for your own explorations, FractalWorks is available in the Mac App store (http://tinyurl.com/fractalworks).

4. Carl R. Spitznagel, "Julia Sets," *Mathematical Vignettes* (blog), 2000, http://www.jcu.edu/math/vignettes/Julia.htm.

5. Pickover, *The Math Book*, 356.

6. Emilio Gomariz, "Scott Draves" (interview), *Triangulation Blog*, January 14, 2011, http://www.triangulationblog.com/2011/01/scott-draves.html.

# resources

For information on fractals and fractal-generating algorithms or software, consult the following books and websites:

## books

Briggs, J. *Fractals: The Patterns of Chaos*. New York, Simon & Schuster, 1992.

Gleick, J., *Chaos: Making a New Science* (revised edition). New York: Penguin Books, 2008.

Lesmoir-Gordon, N., Rood, W., & Edney, R., *Introducing Fractals*. London: Icon, 2009.

Mandelbrot, B., *The Fractal Geometry of Nature*. New York: W. H. Freeman, 1982.

Peitgen, H., & Richter, *The Beauty of Fractals*. Berlin, New York: Springer-Verlag, 1986.

Pickover, C., *The Math Book*. New York: Sterling, 2009.

Pickover, C., *The Pattern Book: Fractals, Art, and Nature*. River Edge, NJ: World Scientific, 1995.

## websites

Apophysis 7X, http://apophysis-7x.org/

Chaoscope, http://www.chaoscope.org/

ChaosPro, http://www.chaospro.de/

Electric Sheep, http://electricsheep.org/

Fractal Foundation, http://fractalfoundation.org/resources/fractal-software/

Mandelbulb 3D, http://www.fractalforums.com/mandelbulb-3d/

Mandelbulber, http://www.mandelbulber.com/

MojoWorld, http://www.pandromeda.com/

Structure Synth, http://structuresynth.sourceforge.net/

# image credits

© **Duncan Champney:** i, iv, 2, 5, 7, 8, 11–15, 16 (bottom), 17 (top), 18, 21–26, 28–31, 53, 75, 78, 79, 134

© **Aleksey Osipenkov (osipenkov.art@gmail.com):** 35 "Alien World 4," 36 "Tunnels of Nibiru," 44 "Keepers of Time," 46 "The Galactic Hologram," 47 "Gravitational Astronomy," 48 "Panspermia," 49 "Living Stones," 50 "At a Depth," 59 "The Emitter of Dreams," 80 "Wormhole," 81 "Io," 82 "Anatomy of a Cyber-brain," 83 "Tower of Shakti," 84 "Setebos," 85 "Floral Dreaming," 86 "Gardens on Nibiru," 87 "Antigravitator," 88 "Exoplanet GJ1214b," 91 "Bunker," 92 "Light," 93 "Weird the Dragon," 95 "Elevator to the Center of Nibiru," 108 "Planet the Puzzle," 113 "Spiderweb," 118 "The Astral Commutator," 120 "Alien World 3," 121 "City in the Astral," 122 "In an Orion Nebula," 142 "Pass to Eden," 143 "Cosmos Labyrinths," 156 "In Distant Space"

© **S. Rathinagiri:** iii, 33, 34, 38–43, 45, 51, 52, 56, 58, 60–62, 66–74, 77, 89, 90, 94, 95–107, 109–112, 114–117, 119, 124–126, 129–133, 136, 137, 139–141, 145–153

© **Fernando Vallejo:** 4, 6, 9, 10, 16 (top), 17 (bottom), 19, 27, 54, 55, 57, 63–65, 76, 123, 127, 128, 135, 138, 144

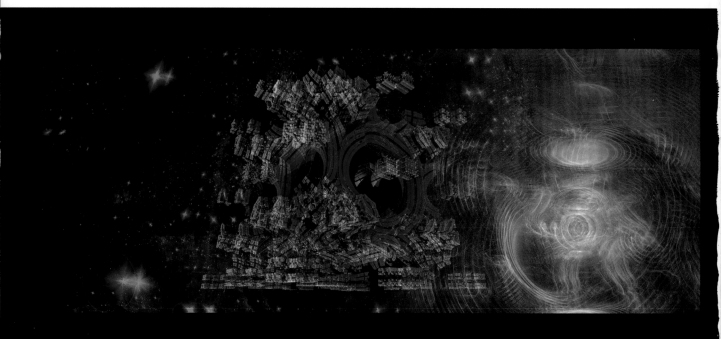